IMPROVING
Garden Soils

WRITER
PHILIP HARDGRAVE

PHOTOGRAPHER
ROBERT FOOTHORAP

ILLUSTRATOR
EDITH ALLGOOD

AVON BOOKS  NEW YORK

Acquisition, Development and Production Services by BMR, of Corte Madera, CA

Acquisition: JACK JENNINGS, BOB DOLEZAL

Series Concept: BOB DOLEZAL

Developmental Editing: BOB DOLEZAL

Photographic Director: ALAN COPELAND

Cover Photo: BARRY SHAPIRO

Interior Art: EDITH ALLGOOD

North American Map: RON HILDEBRAND

Copy Editing: NAOMI LUCKS, JANET REED

Proofreader: TOM HASSETT

Typography and Page Layout: BARBARA GELFAND

Index: JIM NAGEOTTE

Horticulturist and Site Scout: PEGGY HENRY

Color Separations: PREPRESS ASSEMBLY INCORPORATED

Printing and Binding: PENDELL PRINTING INC.

Production Management: THOMAS E. DORSANEO, JANE RYAN

Film: FUJI VELVIA

Additional photo credits: Bob Dolezal, pages 8-9.

First Avon Books Trade Printing: February 1992

ISBN: 0-380-76666-3

Library of Congress Catalog Card Number: 91-67351

Notice: The information contained in this book is true and complete to the best of our knowledge. All recommendations are made without any guarantees on the part of the authors, NK Lawn and Garden Co., or BMR. Because the means, materials and procedures followed by homeowners are beyond our control, the author and publisher disclaim all liability in connection with the use of this information.

AVON BOOKS
A division of
The Hearst Corporation
1350 Avenue of the Americas
New York, New York 10019

AVON TRADEMARK REG. U.S. PAT. OFF. AND IN OTHER COUNTRIES, MARCA REGISTRADA, HECHO EN U.S.A.

91   92   93   94   95   10   9   8   7   6   5   4   3   2   1

IMPROVING GARDEN SOILS

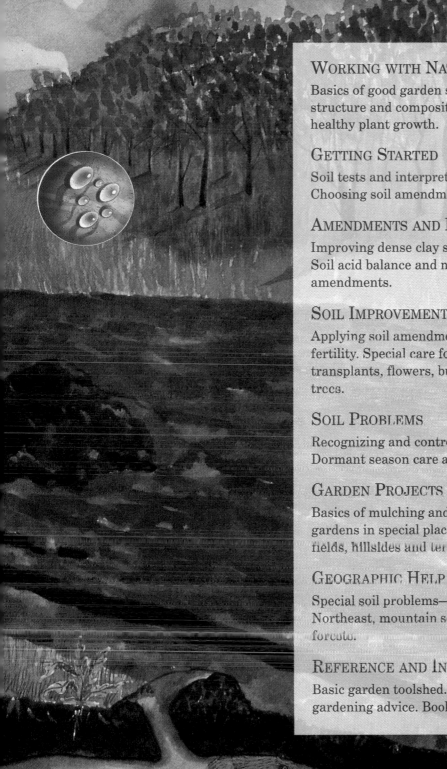

GOOD GARDEN SOIL

You don't have to be a soil scientist to grow healthy plants and vegetables, but a basic knowledge of soil structure and ecology will greatly increase your success and enjoyment as a gardener.

Improving garden soil begins with knowing the native soil of your area. This soil is made up of inorganic matter (rocks and minerals of all kinds) and organic matter (decomposing plant and animal materials), with air and water in between the pore spaces supplying nutrients needed for plant growth.

A well-structured soil is about half mineral and organic matter, a quarter air and a quarter water.

HUMUS
A rich brown or black product of the decomposition of plants, leaves and living organisms.

TOPSOIL
The surface layer of earth moved in cultivation, teeming with plant and animal life.

SUBSOIL
The soil zone beneath the topsoil, which may have to be broken up in order to permit drainage and root development.

SOIL DISTURBANCE
Tunneling by worms, insects and other organisms that helps to break up soil particles.

PLANT ROOTS
Roots carry nutrients from the soil to stems and leaves.

SOIL TYPES

Consult this soil map to determine the general type of soil in your area. In the eastern part of the country, rainfall usually exceeds evaporation, and calcium carbonate is leached from the soil, making it more acidic. In the lower rainfall areas of the west, evaporation exceeds precipitation, and soils become alkaline.

Rock and Ice
Ice or bedrock with little native soil.

Boggy Sediments
Soggy, organic-rich tundra covering permafrost.

Varied Mountain Soil
Varied, coarse soils drain quickly and leach nutrients.

Organic-rich Steppes
Deep, rich black and brown soils, excellent for gardening.

Gray-brown Clays
Light colored, sandy soils leached of water-borne nutrients.

Alkalined Deserts
Low organic content, often with accumulated salts.

Coarse Sands
Light color, sandy structure. Quick to drain.

Southeast Clays
Moist and sticky, low in natural nutrients.

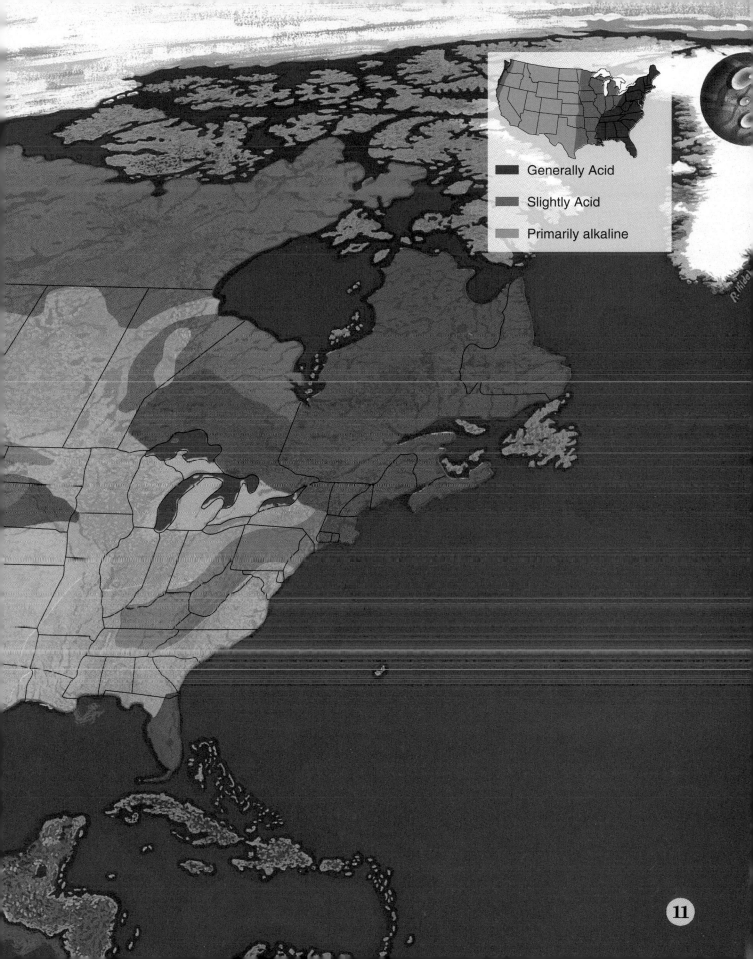

Generally Acid

Slightly Acid

Primarily alkaline

11

GARDEN SOIL BASICS

All garden soils are mixtures of clay, silt, sand and loam. Soils with too much clay are dense, hard to work and drain poorly. Sandy soils are too loose and dry out quickly. The best garden soils—loams and humus—combine all three with decayed plant matter, are easy to work and hold good amounts of water and air.

Silt is larger than clay and smaller than sand, 1/100–1/10,000 of an inch. It retains water well, but lacks important air space between particles.

Clay particles are dustlike, less than 1/10,000 of an inch in diameter. Wet clay is sticky and hard to work. Dry clay is hard, locking out air and water.

Loam contains nearly equal parts of clay, silt and sand. It is the finest, most easily worked soil. It holds water, air and nutrients easily.

Sand has the largest soil particles, 1/25–1/100 of an inch. It contains lots of airspace but holds water and nutrients poorly.

GARDEN SOIL AND PLANT GROWTH

Vigorous plants have good color and deep roots from a balanced soil rich with the basic nutrients—nitrogen, phosphorus and potassium. The foliage is a rich, dark green and the roots run deep, making for heavy fruit bearing.

Abundant flowers

Vigorous foliage

Healthy stem

Strong root system

Nitrogen-poor soil causes yellow-green leaf color and poor stem and leaf growth.

Phosphorus-poor soil can cause purple edges and tips, shallow root systems and slow growth.

Lack of potassium in the soil exposes your plants to disease and keeps nutrients from getting to their roots.

SOIL AND PLANTS

Good soil structure makes a world of difference in how well your plants grow. If your soil is too packed, there isn't enough pore space for air to penetrate and the soil can't draw water up and down to feed root systems. Surface water runs off without penetrating and plant roots grow weak and shallow. If your soil structure is too grainy, nutrients will penetrate quickly but won't stay long enough for roots to use them.

Poor drainage is also a major cause of plant problems. If your water table is too high or your garden has poor drainage, your plants will be vulnerable to root rot, toxic salt buildup and shallow root structures that cause plants to wilt rapidly in dry spells. With poor drainage harmful parasites will flourish but earthworms won't, and little oxygen will reach your plants' roots.

The ideal texture is a loose, loamy soil with the consistency of a handful of bread crumbs. The soil particles should have small, irregular sizes and shapes so that water is easily absorbed, roots get plenty of air and the soil is easy to work. The nutrients in the soil will filter down to a deep root system, and drainage will be excellent at all soil depths.

A loose, loamy soil structure can help avoid the kinds of nutrient deficiencies that make plants give off distress signals—but even these distress signals can be helpful. You can see the real relationship between your plants and your soil as well as many chemical tests can, if you know what each symptom represents. Finding out what nutrients your plants need begins with a close inspection for the early warning signs of soil nutrient deficiency.

BASIC SOIL TESTING

WHAT SOIL TESTS MEASURE

Soil tests measure the amount of nutrients your garden soil can supply to your plants, and tell you which fertilizers you need to add for good growth and high yields. The four basic soil tests are for chemical content—nitrogen, phosphorus, potassium—and pH (acidity or alkalinity).

Nitrogen (N) is needed to produce leaf growth and good chlorophyll, but too much nitrogen can soften plant tissue and retard flower and fruit development. Phosphorus (P) promotes growth of root systems, and stimulates flower and seed formation. Potassium or potash (K) carries carbohydrates throughout the entire plant, building strength and resistance to disease. The pH test measures overall acidity or alkalinity of the soil.

Most plants do best in slightly acid soil, but leaching of limestone and shell deposits tends to deplete natural calcium carbonate from the ground. When this happens, the soil becomes too acid, lacking not only calcium but also the other nutrients that calcium unlocks in the soil.

You should mark the location and planter beds of each soil sample if you are testing a large garden or field. For smaller gardens you can mix several samples together in a single test. Make a written list of the plants being grown in the soil; any fertilizers, lime or sulfur already being used; what kind of drainage and irrigation is being used; and any symptoms of deficiency you may have noticed. Testing services will provide complete soil test results. Your garden store staff will help interpret the results and provide other useful information.

First Dig several holes, 1 ft. round and 1 ft. deep, in various places around the garden.

Then Take a cup of soil from the side of the hole and place it in a clean bucket or other container.

Third Mix the samples together to get an average mixture of soil from all parts of your garden.

Last Take your sample to your Agricultural Extension Service or garden store.

UNDERSTANDING SOIL TEST RESULTS

Whether you perform your own soil tests or send a sample to a nearby Agricultural Extension Service, nursery or college soil lab, you need to be able to properly interpret the results. Test results are usually expressed as the percentages of nitrogen, phosphorus and potassium needed in fertilizer used to correct deficiencies; these percentages are the three numbers found on the labeling of most fertilizers. A complete, all-purpose fertilizer with a guaranteed analysis of 10-8-8, for example, contains 10 percent nitrogen, 8 percent phosphoric acid, and 8 percent water-soluble potash. If your soil tests high in nitrogen, but low in phosphorus and potash, you should select a fertilizer with a lower nitrogen content, such as a 5-10-10 formula. Organic fertilizers such as bone meal and cottonseed meal are usually high in one element and low in the other two, but composted manure can be a well-balanced complete fertilizer, depending upon the animal and diet that produced it.

You can easily correct most nutrient deficiencies by the proper application of fertilizers (see pgs. 24-25), or pH imbalance with application of lime or sulfur (see pgs. 26-27), but the amount and the method of application will vary depending on the needs of your soil and plants. Soil test results for N, P and K translate into the amount of each fertilizer that your soil needs; divide these figures into smaller quantities for smaller gardens. In the same way, pH test results will translate into an amount of lime or sulfur needed to bring the soil back into the desired range. You can actually fine-tune the soil pH to your plants' individual requirements— blueberries and lilies like fairly acid soil, for example, while cabbages and carnations prefer neutral or slightly alkaline soil.

WATER DRAINAGE TESTING

Good drainage is essential for balancing soil aeration and percolation, so you should try to choose a garden location with adequate natural drainage. Poor drainage can be improved by mixing amendments into the soil, digging ditches around the garden or laying drainpipes to carry off excess water.

Testing Drainage

First Dig a 1 ft. round and 2 ft. deep hole in your garden soil.

Then If the soil has received water recently, cover the hole and surrounding soil with plastic and wait 7 days until the soil dries.

QUICK TEST

Take a handful of moderately wet dirt, squeeze it tight and then release. If the soil remains stuck together, or falls right apart, it will not drain properly. An ideal soil will crumble slowly when released.

Third Fill the hole to the brim with water and begin timing.

Last If the water drains in less than 5 minutes, the soil is too loose; if more than 15 minutes, the soil is too dense.

TOO DENSE, TOO LOOSE

Just as the amount of rainfall varies greatly from one region to another, so does the ability of different garden soils to absorb water. Drainage can take place externally, as surface runoff into ditches, streams or drains; or internally, as infiltration down through the soil's root holes, cracks, and pores created by insects, burrowing animals and the soil texture itself. If your soil surface is dry and hard, even a heavy rainfall will run off quickly, leaving little or no water to percolate down into the ground. If your soil is wet and waterlogged, on the other hand, excess water will prevent the air from reaching your plant roots, and the plants will eventually die.

Special drainage problems can occur when the topsoil is caked or mottled, or when the native soil is stripped away to the subsoil, as often happens in excavation or new home construction. The exposed subsoil may be too dry and compacted to permit water infiltration, or it may be so wet from a high water table that pumping or artificial drainage is needed. In most cases you will have to build up this poorly drained ground with soil amendments of the proper structure and texture.

A quick drainage test of your soil will reveal whether you have any such problems, and suggest ways in which soil amendments can build up your topsoil to a suitable condition. Depending on whether you have dense, clay soil (see pgs. 20–21) or loose, sandy soil (see pgs. 22–23), you can choose from among many different soil amendments to improve the workability and drainage of your soil. Dense soil, for example, may require organic material worked into the topsoil, or even drainage holes filled with sand or gravel, to get through the layer of clay blocking infiltration.

IMPROVING DENSE SOILS

HEAVY SOILS

To improve the texture and drainage of dense, clay soils, add almost any material that creates air spaces and breaks up the compacted clay. The most obvious material is sand, but other mineral amendments—such as vermiculite, pumice and gypsum—can be used to better effect. Use organic matter, such as humus, compost or peat moss, to correct poor drainage in clay. Most soil only has around 3–5 percent humus naturally, but you can add humus up to about 25–35 percent of soil volume with good results. The humus will create pore spaces between the packed layers of fine clay particles, acting like a sponge and allowing air and water to penetrate the soil.

Inorganic materials, such as sand, gypsum and vermiculite, can always be used to build up pore space to the ideal ratio of around 50 percent of volume. Unfortunately, the amount of material needed to reach this proportion is considerable—as much as 25–50 percent of the total volume of your finished product.

For example, a 100 cu. ft. volume of heavy clay soil may require an equal amount of sand or gypsum to give it a loose, crumbly texture. The high cost of mineral amendments usually prohibits their use in larger gardens. These materials do have one big advantage—they decompose slowly and remain in the soil for years.

Organic materials, on the other hand, break down fairly rapidly, especially if they are well watered and aerated. Bark, peat moss, leaves and composted material decompose into humus, however, and humus serves as a binding material to form new aggregates from clay particles in the soil. Using well-decayed organic matter is the most cost-effective solution to poorly draining clay soils.

COMPOST
Loosens heavy clay soil, improves drainage and aeration if mixed in thoroughly by hand spading or tilling.

GYPSUM
May help form larger soil
aggregates from clay particles,
lower available sodium levels and
provide supplemental calcium
and sulfur.

HUMUS
Improves soil aeration and water
penetration in clay soils, and
retains water and soluble
nutrients in sandy soils.

ORGANIC PLANTING MIXES
Trench in and turn over so they
total 33–50 percent of the volume
of your native soil.

IMPROVING LOOSE, SANDY SOILS

SUPER-ABSORBENT POLYMERS
Absorb many times their weight in water, and hold moisture for slow release into root systems.

COMPOST
Decayed organic material that gives texture and added nutrients to sandy soils.

ORGANIC PLANTING MIX
Rich in plant nutrients and improves texture.

GROUND BARK
Decomposes slowly. Add nitrogen-rich fertilizer to avoid depletion of soil.

VERMICULITE
Little nutritional benefit but improves aeration and water retention.

IMPROVING FAST-DRAINING SOILS

Sand particles range in size from 1/250th up to 1/12th of an inch, but all loose and sandy soils share the same basic problems. The particles have large pore spaces between them, and water drains through them so quickly that the soil dries up before moisture and water-borne nutrients can be absorbed by plant roots. A light, sandy soil might have 70 percent sand, 20 percent silt and 10 percent clay. Adding clay soil is one way to build up sandy soil to the ideal balance of equal parts sand, clay and silt. Few of us have both kinds of soil in the same garden. Soil amendments are usually required to build sandy soil into a well-balanced garden soil.

Many of the same materials used to improve sandy soils are also effective with clay soils, but for different reasons. Organic materials always help increase moisture retention and aeration, but they can speed up the decomposition of humus and its nutrients. Avoid use of too much slower-decaying materials such as ground-up bark, peat moss or sawdust in your sandy soil. As they decompose, they will consume most of the soil's nitrogen. Always compensate by adding nitrogen fertilizer regularly.

Mineral amendments such as pumice and vermiculite are best suited to these sandy soils, but their high cost may rule out use for large gardens. Remember, soil amendments will make up as much as 50 percent of your final soil volume, so choosing the right soil amendment can be as much a question of economy as effectiveness.

A new technology of chemical water-holding agents, called "super-absorbent polymers," can dramatically increase water retention in soils with too much drainage. These water-holding polymers are expensive, and are best used in quick-draining potting and container soils.

CORRECTING SOIL ACID BALANCE

Good watering practices prevent acid or salt buildup in garden soil. Water heavily but infrequently, and test the pH of your soil and water.

ABOUT BALANCED SOIL

Soil pH measures the acid-alkaline balance, on a scale of 0 (acid) to 14 (alkaline), with pH 7.0 being neutral. Most plants prefer a pH somewhere around neutral, and any pH outside a range of 4–8 indicates a severe imbalance. A low pH soil will lack calcium, magnesium and phosphorus. A high pH soil, on the other hand, indicates excess salt, sodium or calcium carbonate, which can cause leaf burning and yellowing.

Acid soils usually occur in high rainfall areas, or in sandy, wet organic soils. Most of the East has naturally acid soil, but so do rainy areas along the Pacific coast. Soils of the drier parts of the West are usually alkaline, often leaving surface salt deposits. Your soil's acidity or alkalinity directly affects the availability of plant nutrients. Most nutrients are locked up in highly alkaline soils, and only iron is freely available in highly acid soils.

Each plant has its own range of pH tolerance, because different plants have different nutrient needs. For example, spinach and asparagus have a high salt tolerance, while corn and celery do not. Apply lime to correct acidity, or garden sulfur to correct alkalinity. Once the lime or sulfur is leached out of the soil, the pH imbalance will return. Repeat as necessary.

Farmers measure applications in tons per acre, but gardeners use these rules of thumb: add about 4 lbs. of lime per 100 sq. ft. for every pH point below 6.5, and 1 lb. of sulfur per 100 sq. ft.

Sulfur

PLANTS THAT PREFER VERY ACID SOIL (pH 4.0 TO 6.0)

Azalea

Blueberry

Lily

Marigold

Peanut

Potato

Raspberry

Rhododendron

Watermelon

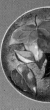

Lime

**PLANTS THAT PREFER
SLIGHTLY ACID SOIL
(pH 6.0 TO 6.5)**

Apple

Cherry

Corn

Eggplant

Kale

Parsley

Pea

Pepper

Squash

Tomato

**PLANTS THAT PREFER
SLIGHTLY ALKALINE SOIL
(pH 7.0 TO 8.0)**

Asparagus

Broccoli

Cabbage

Cucumber

Lettuce

Muskmelon

Onion

Spinach

25

Correcting Nutrient Deficiencies

Choosing a Fertilizer

There are many different kinds of fertilizers, and gardeners are always debating the merits of organic versus synthetic, dry versus liquid. Each will correct soil nutrient deficiencies.

Synthetic fertilizers have the advantage of pre-mixed proportions, compact storage and low cost. However, they may burn plants. Always apply according to label directions.

Organic fertilizers are usually bulkier and may be more difficult to apply—or more expensive—than synthetics. Organics include blood and bone meal, composted manure, fish emulsion, soybean meal and wood ashes.

Each fertilizer has a different amount of nutrients. Read the label to see how much nitrogen, phosphorus and potassium your fertilizer contains. The percent by weight appears as a series of three numbers, in order, for nitrogen (N), phosphorus (P) and potassium (K). The label also shows the recommended rate to apply.

Nitrogen dissolves in water and is removed from soil by erosion and leaching. Because of this, it needs replacement more often than phosphorus or potassium.

FISH MEAL
Nitrogen-rich fish remains have lower proportions of phosphorus and potash.

BONE MEAL
Very rich in phosphorus but low in potassium, bone meal is good for bare-root planting.

ORGANIC ™
LAWN FERTILIZER
10-3-4   FEEDS 2,500 SQ. FT.

NATURE'S CHOICE LAWN FERTILIZER
COMPLEMENTS YOUR LAWN'S NATURAL LIFE CYCLE

Nutrients are delivered in a formula that feeds your lawn in the proportion grasses require:
- Nitrogen for vigorous, green growth
- Phosphorus for strong, deep roots
- Potassium for resilience and durability

Vigorous root systems promote a lush, thick, dense turf that chokes out weeds before they become established.

Natural microorganisms in the soil break down organic matter, which improves tilth and reduces thatch, where fungal diseases can breed.

LIQUID FERTILIZERS

Liquid fertilizers contain their dissolved nutrients in solution and are diluted with water. Chemical burning is more easily avoided than with dry fertilizers. Liquids and emulsions are not well suited for larger gardens, however, since they are more expensive than dry mixes and tend to leach out of the surface soil more rapidly. Smaller gardens, planter boxes and containers are better places for liquid fertilizers, which can be applied locally in formulas adjusted to individual plants.

FISH EMULSION
High in nitrogen, can be sprayed as a diluted liquid.

SYNTHETIC GRANULATED MIXES
Pre-mixed nutrients are easy to use but can cause fertilizer burn if improperly applied.

MANURE
Mild, well-balanced and a good soil conditioner if allowed to decompose.

CHOOSING PURCHASED AMENDMENTS

SELECTING AMENDMENTS

Once you have determined your soil's texture and composition, you will know what to look for in a soil amendment.

Organic amendments tend to be short-term agents that need regular replacement, so look for bulk prices on larger quantities, especially if you have a large area to cover. Even small garden beds and planters require a surprisingly large amount of amendment material to really improve texture and drainage, so you may want to consider longer-term organic agents such as ground bark or peat moss. Lightweight amendments such as vermiculite are much less expensive when purchased in big 4–5 cu. ft. bales rather than in small bags, but you may have to do some hunting to find commercial quantities.

The quantity of amendment needed for your garden project will depend on the percentage of new soil to be mixed in with the native soil. Most organic amendments have to be mixed in ratios of 25–50 percent of the old soil volume in order to be effective. Lighter, more persistent amendments such as vermiculite and peat moss only require adding 10–20 percent by volume (bearing in mind that peat moss expands when loosened from the bale). Application rates can be figured in terms of inches of coverage, pounds per 100 sq. ft., or cubic feet or yards per 100 sq. ft.

You shouldn't hesitate to mix in as much material as you need. If you are at all uncertain, follow the same advice with amendments as for fertilizers. Mix in a little bit at a time—if you add too much amendment you'll have to go back and add more soil later.

APPLICATION RATES FOR VARIOUS AMENDMENTS

ORGANIC MATERIAL (COMPOST, HUMUS) PER 100 SQ. FT.

Depth of Coverage (Inches)	Percent of Soil Volume		Amount of Amendment	
	At 12 in.	At 18 in.	Cu. Yds.	Cu. Ft.
1	8%	5.5%	1/3	9
3	25%	16%	1	27
4	33%	25%	1 1/3	33
6	50%	33%	2	50

LIGHT AMENDMENTS (VERMICULITE, PERLITE, PEAT MOSS) FOR 25 SQ. FT.

Percent of Soil by Volume	Amount of Amendment (cu. ft.)	
	For 12 in.	For 18 in.
10%	2.5	3.75
20%	5.0	7.5

APPLYING AMENDMENTS TO LARGE AREAS

Carry your soil amendment to larger sites in a garden cart or wheelbarrow. Spread the material evenly, hand spading into the soil, double-digging for better mixing. Avoid tilling wet soil, and be careful not to over-till dry soil to a powder when using a power rotary tiller.

Application of Amendments

First Determine amendment coverage rate, purchase and bring to garden site.

Or Apply thinner coats of amendment and work material into soil.

Then Spread over the soil bed, rake level and measure coverage depth.

Next Apply fertilizer, lime or sulfur with drop-spreader.

Either Top dress the old soil with a thick, balanced layer of new topsoil mix.

Last Work into soil at right angle to original tilling direction.

IMPROVING SOILS FOR LAWNS AND GROUNDCOVERS

LAWN SOIL PROBLEMS

Growing rich, green turf or groundcover is almost impossible without adequate soil preparation. Some lawn soil problems, such as hardpan, are natural, but many are manmade. Footpaths and trails, improper preparation for planting and surface scalping in construction leave soil dense and compacted. Given well-textured soil, most lawn grasses will develop hearty root systems—10–60 in. deep. They need water and air to thrive.

Aerify your lawn when the ground is moist enough to allow easy penetration. Aerification can be done with hand tools, but rented power aerators do a better job.

Thatch is a thick, yellow mat of old grass between the blades and the soil surface. When it is more than 1/2 in. thick, a rented dethatcher or dethatching rake will remove it. Power dethatchers save time and energy if your lawn area is large. Only dethatch and aerify your lawn while its grass is actively growing.

Apply soil amendments and fertilizers after dethatching and aerification. Ideal lawn soil is a sandy loam with nearly equal parts of sand, clay, silt and decomposed plant matter. Half of its volume should be pore spaces.

Turfgrass soils should be pH 6.5–7.0, while groundcovers prefer soil of pH 6.0–6.5. Correct acid imbalances with garden lime, alkalinity with sulfur. Apply fertilizers in early spring and every 4–6 weeks throughout the growing season.

Correcting Compacted Soils

First Dense, compacted soil causes shallow root growth. Turf dies in warm weather.

Third Rake and remove plugs, or let them dry and decompose into the lawn.

Then Aerification machine pulls up soil plugs. Water and air penetrates.

Last Fill in any bare spots by reseeding with seed and topsoil mix.

Dissolve plugs from aerification into lawn with watering

TRANSPLANTS AND SOIL

CONTAINER-GROWN PLANTS

Nursery-grown plants should be removed from their containers carefully to prevent damaging their stems and roots. Never pull up—always turn the plant upside down. The root ball will usually come out of plastic or peat pots easily, but tapered metal cans may have to be cut away with shears or a can-cutter. Mix a dry, high-phosphate fertilizer with the soil from the hole. About 1–2 tsps. for 1- or 2-gal. containers, or 1/4 cup for 5-gal. containers, is right.

Before placing the plant in the hole, trim back any excess or damaged foliage and score cuts through any encircling or matted roots. Fill the base of the hole 1/4 full with soil. Carefully set the plant down into the hole, supporting its weight by holding the rootball. Make sure that the top of the rootball is even with or slightly above the soil surface, then fill in the sides of the hole with more soil.

For larger shrubs and trees, use the rest of the backfill to build a watering basin around the outer edge of the filled hole. Slope the soil away from the stem, so that water will form a ring around the outer edge of the roots. Soak the soil thoroughly and make any final adjustments to the plant's position. If the top of the rootball is too low, now is the time to take out the plant and add soil to the base of the hole.

Plant and fertilize bare-root shrubs and trees during their dormant period. When planting bare roots, be sure to work the backfilled soil in between the roots to eliminate any air pockets.

Transplanting Container-Grown Plants

First Dig a hole half again as wide and deep as the container, setting aside soil from the hole.

Third Prune excess or damaged foliage, cut encircling and matted roots with a knife or trowel.

Next Fill hole 1/4 full and tamp. Support plant by root ball, set in hole, adjust height and fill in sides.

Then Loosen plant from container, invert while supporting the rootball and tap out with palm of hand.

Fourth Add high phosphate fertilizer to the native soil mix.

Last Water thoroughly until bubbling stops. Add soil as it settles.

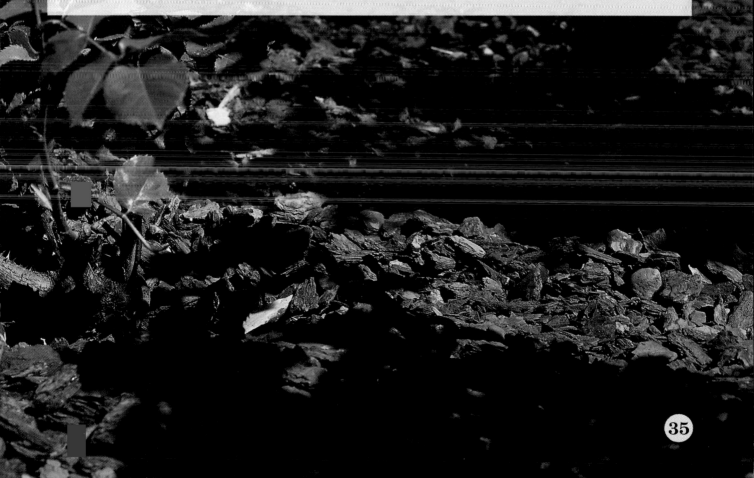

Preparing a Shrub Landscape

First Conduct a soil test (see pg. 16) to learn your soil's nutrient and pH needs.

Fourth When planting, use a high-phosphorus fertilizer to promote root growth.

Then Test drainage (see pg. 18) to find if soil structure is too dense or coarse.

Next Build a watering basin around each plant, just outside the line of its foliage.

Third Amend the soil before planting by working in compost and nitrogen-rich fertilizers.

Last Always mulch the soil surface, but keep mulch away from the plant's trunk.

SHRUBS AND SOILS

Most landscape shrubs are hardy plants that adapt well to a wide range of conditions, but leaching and breakdown of organic matter can lead to soil problems, including compaction, alkaline soil buildup (pH imbalance) and nitrogen deficiency. If your soil has some or all of these problems, correct them before planting your shrubs.

Alkaline buildup also causes chlorosis, a deficiency that causes leaf yellowing. Azaleas, gardenias, laurel, rhododendrons and other shrubs that prefer an acid soil of pH 6.0 or lower may develop chlorosis in alkaline or even neutral soils where hard groundwater is a problem. Correct this condition by feeding with a chelated iron fertilizer that keeps soluble iron, copper, zinc and other micronutrients available.

Acid balance is another problem, especially if the plants are not suited to your soil's acid or alkaline condition. Most shrubs prefer soil pH between 5.5 and 7.5, but some desert shrubs will thrive in soils as high as 8.0 to 9.0.

To correct alkaline soil conditions in general, use garden sulfur or ferrous sulfate (see pgs. 24–25) in the amounts indicated by your soil test. Acidic soils may be corrected by adding garden lime—never use quicklime.

Water deeply and improve shrub soils before planting by adding compost, humus, well-rotted manure or nitrogen-stabilized wood bark, chips or sawdust. These organic amendments will help hold water around the root system, permit air to penetrate and improve the soil's texture so that shrubs become well rooted. Build a basin of soil around the drip line of the shrub for easy watering, and add organic mulch over the bare soil. Take care not to apply the mulch directly to the exposed roots or trunk of the shrub.

SOIL CARE AND DECIDUOUS TREES

Amending Soil for Landscape Trees

First Conduct a soil test (see pg. 16) for needed nutrients and pH balance.

Third Hole should be twice as deep and wide as rootball.

Then Test that soil drains properly (see pg. 18).

Last Mix fertilizer with all native soil. Amend soil texture only under rootball.

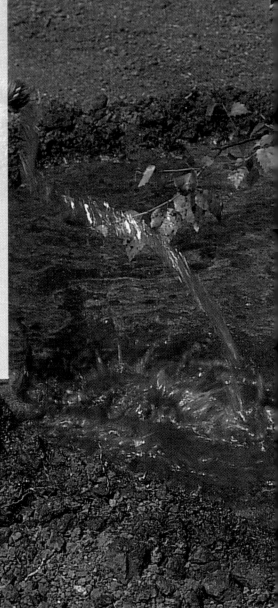

TREE SOIL PROBLEMS

Trees come in all sizes, but they are usually the largest plants in the landscape garden. It might seem that landscape trees should grow well in any native soil, but in fact trees require proper nutrients, acid balance and maintenance just like other garden plants. Tree root systems can reach down into the subsoil all the way to bedrock; but if the soil is too dense, roots will run sideways and back to the surface in search of air and water.

Most trees prefer heavy, deep watering; lighter, frequent watering causes shallow root systems and stunted growth. Younger trees should be watered frequently until root systems are well developed. The quick soil drainage test (see pgs. 18–19) will determine whether your trees are getting enough water, but make sure you take your tree soil sample from at least one foot deep in the ground.

Trees often compete for nutrients with shrubs and other trees, particularly if large trees are planted too close to each other. Although it is difficult to do much about nutrient competition deep underground, regular watering and fertilization will usually help an undernourished tree. Long, steady watering with a soaker hose and periodic watering with a deep root irrigator will make sure the tree is not deprived of water below ground. Aeration with a pitchfork or aerator tool also helps air and water get to the root system.

Feeding an established tree can be done with liquid, solid or granular fertilizers, applied as spray mixes, granules, slow-release spikes or ground pellets. The same rules for testing and correcting for nutrient deficiencies in gardens (see pgs. 26–27) apply to tree fertilization. Use added care not to burn bark or roots by direct application of chemicals.

Renovating a Flower Bed

First Dig perennials, bulbs and corms. Set aside perennials to replant later.

Third Apply amendments and fertilizers to flower bed, dig it in at least 8 in. deep.

Then Cultivate soil 12–16 in. to loosen compacted soil.

Last Plant annuals or replant bulbs, corms and rootstock in freshly worked bed and mulch.

SOIL FOR FLOWERS

Flowers and flowering bulbs have special nutrient requirements. Improving your flower beds' soil will help all ornamentals produce beautiful blossoms, deep roots and healthy foliage. Not all flowering plants have the same maintenance needs, however. Some hybrid roses require constant pruning and feeding, while marigolds, daisies and hardy perennials may need little encouragement.

Good soil conditions for flower beds are the same as for most other plants. Ideal soil has a pH between 5.5 and 7.0, loamy texture and plenty of decomposed organic nutrients. Sandy soil drains too quickly, wilting plant leaves; clay soil holds too much water, drowning plant roots.

Most annuals have shallow roots about 6 in. deep, while perennials may root down a foot or more. For perennials, double-dig the flower bed to a depth of 2 ft., then mix in well-rotted manure, peat and a handful of 0–20–0 superphosphate fertilizer for every 4 cu. ft. of soil volume. Annuals prefer a slightly richer soil than perennials, although individual flowers vary. Marigolds and zinnias thrive in rich organic soil mixes, while portulaca and nasturtiums do well in poorer soils.

Bulbs need loose, porous soil dug 12–16 in. deep. Every bulb has a different proper planting depth—the later you set out your bulbs, the shallower they should be planted. Organic amendments—peat moss, humus and garden compost—will loosen heavy clay soils and help sandy soils hold water. Add bonemeal and superphosphate to the planting hole before setting out bulbs.

Prepare flower beds in the fall for spring planting, and in summer for fall planting. Whenever your garden soil becomes compacted, or depleted of basic nutrients, you need to renovate the flower bed.

VEGETABLE GARDEN SOILS

Garden Soil Preparation

First Spread compost and other soil amendments 2–4 in. deep across the bed, add superphosphate 2 1/2 lbs. per 100 sq. ft.

Third Apply 2–4 in. of compost and fertilizer. Work it in at right angles to trenches.

Next Double-dig first trench two spade lengths deep. Dig second row, turn its topsoil into the first. Repeat for length of garden. Soil from last trench fills first.

Last Level the soil, breaking up or removing any large clods. Rake surface smooth for planting.

SPECIAL NEEDS OF VEGETABLE GARDEN SOILS

Vegetables need all of the same soil nutrients as other garden plants, but more of them. You can correct soil nutrient deficiencies separately for each vegetable crop, or apply fertilizer to individual plants as needed. Nitrogen is depleted from soil more quickly than phosphorus or potash, and needs to be replaced more often to prevent yellowing and stunted growth. Well-rotted manure is an excellent source of nitrogen; allow fresh manure about 1 year to decompose before planting. Phosphorus and potassium deficiencies should also be corrected when preparing soil for planting. Most micronutrients are already present in the soil, but can be added in premixed formulas if necessary.

Some vegetables can convert nitrogen in the air to their own use. Peas, beans, and other members of the legume family have colonies of nitrogen-fixing bacteria on their roots. These bacteria enrich soil by adding nitrates, so planting legumes in various parts of your vegetable garden is one way to keep nitrogen in good supply throughout the growing season.

Soil preparation is all-important for vegetable gardens. Water the ground well several days before planting, then loosen up the soil with a spade or digging fork to a depth of 8–12 in. Double-digging—digging one spade length deep before loosening up the subsoil below with a digging fork—will enable you to work the soil loose as deep as 24 in. More intensive double-digging requires digging a first trench 24 in. deep, removing the soil, then digging a second trench alongside and moving its topsoil into the bottom of the first trench. The subsoil of the second trench is then mixed with humus or other organic material to form the topsoil in the first trench.

SOILS FOR CONTAINER GARDENS

CONTAINER PLANTS

Gardening in boxes or pots creates special needs for watering, drainage and feeding. Ordinary soil is usually too heavy for containers. It compacts quickly and drains slowly. Use porous mixes of potting soil, vermiculite, perlite, peat moss or sand. Plants will require regular watering to restore moisture lost through drainage. They will still need a proper diet of fertilizers and other nutrients to be vigorous and develop good root systems.

Synthetic soils are usually rich in added nutrients, sterile and loose-textured to provide good drainage and easy root penetration. Many modern potting soils actually contain no natural material. They supply their necessary nutrients through additives. "Natural" soil mixes, on the other hand, rely upon organic material such as humus, peat moss, sand and dolomite limestone to provide good texture and fertilization.

Organic gardeners naturally prefer to use real garden soil, but its use may introduce weed seeds, fungal diseases and garden pests at the same time. You can sterilize soil in an oven at 275–325° F for an hour or more, but this also kills off many beneficial microorganisms. With time and experience, one mix or the other will prove itself.

Regardless of the potting mix you choose, soil depletion is a problem in container gardening. Periodically repotting plants into larger containers avoids the loss of nutrients and encourages healthy new growth. Repot whenever the plant roots fill the pot or planter, but before they mat or become rootbound.

Repotting Plants to Larger Containers

Begin Remove plant and rootball from container by tapping on bottom.

Fourth Mix 1 tbsp. 10–10–10 liquid fertilizer, fish emulsion or bone meal into 1 gal. of water.

Then Cut plant foliage back by 1/4 and shape. Cut entwined roots.

Next Fill bottom of planter with mix. Carefully position plant and fill.

Third Mix equal parts of potting soil, utility sand and compost or humus.

Last Water and drain several times. Add soil to replace any lost by settling.

Wood lice

Snails

INSECT, WORM AND MOLLUSK PESTS

Identifying and controlling soil pests is a study in itself, since almost every major animal group has members that live in the ground. Most soil insect pests are actually the grub or pupa of insects that later become airborne, such as flies, cicadas and moths. Other insects such as ants and subsurface springtails spend most of their lives on or below the soil surface.

Worm "pests" differ greatly in their impact. Earthworms feed on dead plant matter and recycle minerals and nutrients upward through the soil, providing plants with nutrients from their waste. Nematodes or roundworms, on the other hand, are parasites, and those that don't infect animals often spend their entire lives inside plants, feeding from within them and laying eggs that can remain dormant for years before hatching and starting the cycle all over again.

Snails and slugs have few of the beneficial qualities of earthworms, since they only tunnel through loose soil and leaf litter. These nocturnal members of the mollusk family are mostly vegetarian, and will eat as much of your plants as they can consume, producing enough waste matter and mucus to form a sticky layer of slime that holds loose soil together. Unfortunately, mollusks prefer a loose, moist surface soil, so any benefits from their movement are far outweighed by the damage they can do to your plants.

Springtails

Earthworms

Slugs

Earwigs

Nematodes

INSECT PEST CONTROL

The solution to most insect pest problems lies in good soil cultivation habits, especially during the crucial months of late fall, winter and early spring, when many soil pests are still in their pupa or grub stage. Like snails and slugs, most insect soil pests prefer moist conditions and loose, leafy soil surfaces. Centipedes, millipedes, wood lice and scorpions can all be controlled effectively by turning over soil and exposing it to air regularly. Wingless insects such as springtails and silverfish spend their lives underground, eating dead plant matter and leaving most healthy plants alone. Because they also serve as food for beetles and flies, however, they cannot be considered entirely harmless. Keeping the soil surface clean and free of clippings, leaves and litter will help control these pests.

Termites and ants that are beneficial in forest soil are usually destructive in man's environment. Other insects, such as earwigs, are annoying in the house but terribly destructive in the garden. Controlling pest infestations becomes more difficult the longer they are ignored. Start by identifying an outbreak. Regularly inspect your garden soil and plants for signs of attack. Dig up a few spades of soil and examine it closely for larvae, grubs and eggs.

If signs of soil pests are found, their damage may usually be limited by cultivating the soil so that air and water may more easily penetrate. If the infestation is ignored or allowed to develop, you may have to rely on more drastic controls, such as spraying, spreading or dusting with either naturally occurring or synthetic insecticides.

SOIL DISEASES

DISEASES IN SOIL

Plant diseases that originate in the soil stem from a wide range of microorganisms, including fungi, bacteria and viruses. There are many microorganisms living in the soil, and not all of them are harmful to plants. Tiny protozoa sometimes feed on bacteria, which are needed to increase soil fertility. Soil fungi are not as numerous as bacteria, but are larger. In highly acid soils, fungi can outnumber bacteria.

Fungi break down dead plant matter and feed on nematodes and amoeba. Fungi also attack living, healthy plants and become a nuisance.

Most soil bacteria are beneficial, decomposing organic material, humus and minerals to form aggregate particles. Nitrogen-fixing bacteria actually benefit plants and soil by converting nitrogen in the air to forms used by plants. Other bacteria, however, are parasitic, causing diseases such as potato scab and sweet potato soil rot.

The tiniest of the soil microorganisms are viruses. Viruses are inanimate parasites that come to life only in a host animal or plant, invading and killing nitrogen-fixing bacteria and fungi. Some viruses are spread by insect pests: aphids carry spinach virus, for example. Spraying the aphids will control the virus.

The best protection against fungus and virus is good drainage, careful weeding and air circulation between and under plant foliage. If your plants show signs of viral or fungal diseases, contact an expert at the Agricultural Extension Service in your area for advice and remedies.

Black rot fungus

Red stele fungus

Blue mold

48

Brown–rot fungus

Damping off

Mosaic virus

49

OTHER SOIL PESTS

Dog

Gopher

Squirrel

Mice

Crow

Mole

ANIMALS AND BIRDS

The small, furry animals that create problems for the gardener are often easy to spot, but difficult to eliminate. Two major burrowing animals are moles and pocket gophers, which live full-time in the ground. Woodchucks, snakes, ground squirrels, prairie dogs, chipmunks and mice also dig into the soil and occasionally damage gardens.

Eliminating food sources for burrowing pests can be difficult, since moles feed mostly on worms and insects, and gophers feed on roots and tubers. Common controls for moles and gophers include traps, gas or poison bait, but these should be used with caution. Flooding is probably the ideal solution, but it is only practical on level ground. Wire mesh planting baskets and mesh netting underneath garden beds will frustrate gophers. Where they are present, avoid using mulch, which offers them good hiding places.

Above ground, squirrels, snakes, mice and other small ground creatures can all be problems in the garden, and chicken wire or nylon netting barriers should be used to prevent them from reaching plants in the first place. Since they can't burrow very far underground, any surface rodents that make it through the barrier are easy prey for traps.

Birds can be most difficult pests, and you may find that the only way to prevent damage to fruit is the use of a broad-mesh net. For ground-level protection of newly planted seedlings and vegetables, plastic row covers are helpful, but you may have to use chicken wire against crows. Most birds are beneficial in reducing rodents and other pests, however. Since they're never fooled for long by scarecrows, it's a good idea to learn to live with them.

DORMANT SEASON SOIL CARE

WINTER GARDEN CARE

Cleaning up your garden's soil surface and cutting back dead foliage on dormant plants will deny disease organisms shelter, and reduce plant debris available to soil pests when springtime comes. Remove all dead leaves, decayed mulch and twigs from the ground between plants, giving special attention to any plants with diseases that thrive in cold, wet soil.

In harsh winter climates where temperatures fall below 10°F, winterize roses and other delicate ornamentals with mounds of soil or straw, being sure to remove any leaves that could harbor diseases. Lay mesh screen over bulbs placed in the ground to prevent animal pests from digging them up. Larger evergreen shrubs and trees will benefit from late autumn spraying with antidessicant oils, which coat them and hold in moisture. Wrapping tree trunks in burlap is another good way to prevent frost damage.

Plant shelters can be built with many types of materials. In autumn, when the first hint of frost arrives, harvest ripe vegetables and cover any immature fruit with plastic, burlap sacking or blankets. Any overwinter vegetables or perennials remaining in the ground should receive a final watering before the frost hits, so that they won't dry out in the frozen ground. If your garden is bare for the winter, mulching with hay or manure will provide a good blanket for the soil. If you have a large garden, you may want to plant a cover crop of ryegrass to prevent soil erosion in bare earth, tilling it back into the soil in spring.

Autumn Care

First Remove all dead leaves, plant debris and limbs from the soil surface.

Third Prune trees and shrubs. In harsh winter climates, protect sensitive plants.

Then Remove and destroy decayed mulch from around plants, shrubs and trees. Replace with new mulch material.

Last Spray plant surfaces with winter dormant oil spray to kill insect pests.

Applying Landscape Mulching

First Prepare and amend the garden soil, set out landscape plants.

Next Clear away leaves, plant debris and litter from soil surface. Cultivate and rake smooth.

Peat moss

Chipped bark

Straw

Third Lay down porous mulch cloth to block weeds, cut holes for plant stems.

Last Spread mulch over cloth in a 2–3 in. layer, level with rake.

Crushed lava

Shredded bark

MULCH AND SOIL

A mulch can be any number of materials— from straw and hay to bark, peat moss and plastic film—that can be easily applied to protect the soil surface. The benefits of mulching are several. Mulch retains moisture during hot, dry summer months, and repels moisture during periods of heavy rain. Mulching helps control weed growth in the soil between the plants, and keeps the soil surface cooler in the summer and warmer in the winter by insulating it from the elements.

Some gardeners lay down organic mulch, bark or peat moss just because they like its appearance, but the best reasons for mulching are to speed up plant growth and improve yields. The most effective mulches are probably the synthetics, such as polyethylene film, nonwoven polyester fabric, and perforated or woven plastic, all of which serve as water vapor barriers and insulators. Synthetic mulching produces particularly rapid growth in warm-weather plants such as tomatoes and strawberries. The only problem with synthetic mulching is that the materials are often unattractive, but their appearance can be easily improved by adding a thin layer of organic mulching material over the manmade material.

Inorganic mulches such as gravel, crushed lava rock or fine pebbles can serve both a cosmetic and functional purpose, but they can also overheat in hot climates. For that matter, solid organic mulches such as bark and peat moss tend to soak up water and develop mold or fungus on their soil side. Most synthetic mulches will last longer than organic mulches, but they tend to tear and disintegrate toward the end of their life. The correct choice of mulch material will depend on the climate, the season, the soil and, not least of all, the individual tastes of the gardener.

MAKING AND USING COMPOST

Adding easy-to-make compost to your garden will help any soil imbalances, since organic compost neutralizes excess acid and alkaline soil. Compost can also be a useful soil amendment, improving texture, aeration and percolation. It also serves as a good mulch material.

MATERIALS

4 x 4 in. posts

2 x 6 in. sideboards

1 x 1 in. slot boards

Posthole cement

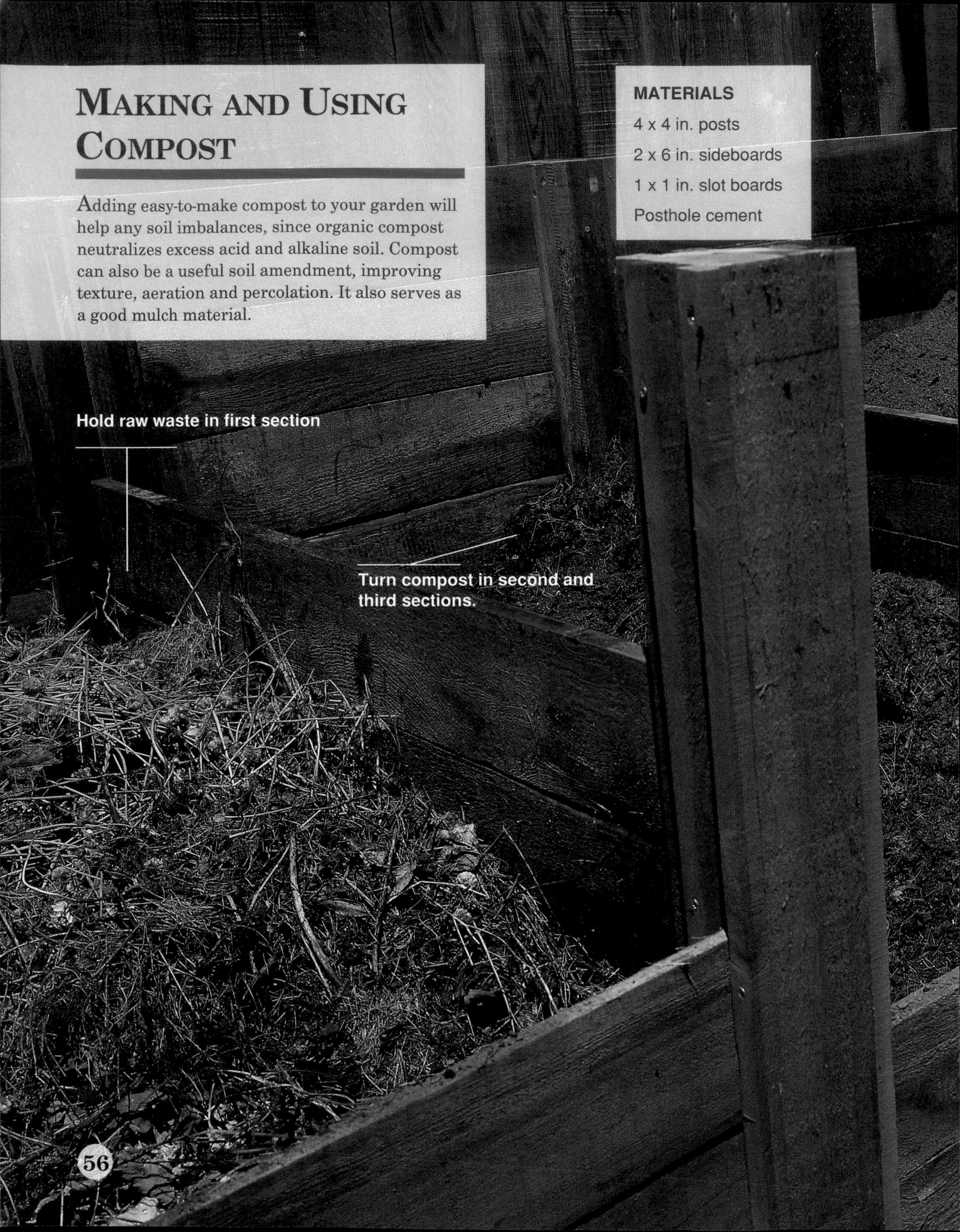

Hold raw waste in first section

Turn compost in second and third sections.

Finished, ready-to-use compost.

3-Bin Compost Box

First Assemble construction materials.

Then After digging postholes 18–24 in. deep, set uprights with posthole cement.

Next Fit the horizontal sideboards into the uprights.

Last Use vertical bin spacers to prevent spillage.

USING A COMPOST PILE

Although compost doesn't necessarily add many nutrients of its own, it does loosen native soil and free up its nutrients, and it improves soil texture.

Compost can be made from many organic materials—tree leaves, grass clippings, dead plants, sawdust, kitchen waste, even rotten fruit. Certain materials should be avoided: tree leaves that are highly acid (oak leaves and pine needles) and hay that contains live seeds or sprigs. Anything toxic or diseased should be left out.

Making a compost pile is the simple but time-consuming process of building up layers of organic materials on top of each other, watering the pile and turning it regularly. Natural bacteria and earthworms in the pile soon go to work, heating it up to between 140° and 160°F and decomposing the ingredients into a rich, dark brown substance with the smell of freshly turned humus.

Aeration and watering will ensure that the compost "cooks," but frequent turning will speed up the process. The fastest decomposition occurs with the addition of nitrogen and the use of mechanical turning devices such as drums. A large pile of between 200 and 500 cu. ft. turned every 2 or 3 days will ripen within a few months. Just 1 cu. ft. of compost begins as 4 or 5 cu. ft. of refuse, so build your compost pile to fit your garden size.

Follow these simple instructions to build and use a garden compost pile.

SPECIAL PROBLEM
RAISED-BED GARDENS

Building a raised-bed garden is a simple project that can give you a highly productive space for flowers and vegetables. Unlike smaller planter boxes, the raised-bed garden is designed to improve poorly drained soil by giving you total control of your soil materials.

MATERIALS

String, stakes for marking (Or use garden lime)

Spade or shovel

Garden tiller (Optional)

Wheelbarrow

2 x 12 green redwood or cedar lumber (Linear feet of bed perimeter)

1 x 3 2-ft. redwood or cedar stakes (2 per corner plus 1 each 16 in.)

1/4 x 3 lag screws (2 per stake plus 3 per corner)

CAUTION

Do not use treated lumber for planters and raised beds if vegetables are to be grown.

First Choose and mark site. Remove any weeds or turf.

Then Till soil. Dig shallow trench along border.

Next Fasten edge boards to stakes with lag screws. Bottom is 2 in. under soil surface.

Last Fill bed with well-rotted compost and added fertilizer. Mix thoroughly.

AVOIDING SOIL PROBLEMS

If your native soil has excess alkalinity or poor drainage, and soil amendment seems not to be working, you may want to consider building a raised-bed garden. Elevating the garden bed will enable you to bring in all new topsoil, increase cultivation depth, and improve drainage. You can't use your regular garden soil, however, since the soil in a raised bed drains and dries out more quickly, leaching out soil nutrients. Build up your topsoil with at least 50 percent organic material, sifted compost or peat moss. Mulching will also help retain moisture, and the raised bed will warm up earlier than the ground around it.

Raised beds are also a simple solution for hardpan layers which obstruct drainage and prevent large plants from taking root deeply enough If your garden site is low-lying and floods continually, tile or tube drainage may be needed (see pgs. 60–61). Otherwise, a raised bed can improve drainage so effectively that water retention actually becomes a problem. Raised-bed gardens have to be watered so heavily that they may be inadvisable in very hot climates, where evaporation occurs quickly.

Building raised beds can be as simple as joining together 2 x 6 in. untreated lumber into rectangular boxes, and setting them down side by side with paths in between. Leave enough space to roll a wheelbarrow between the beds; paths 2–3 ft. wide covered with shredded bark or gravel are both functional and attractive. Use part of the soil dug from these pathways to fill the beds, mixing this native soil with the amendments appropriate to your soil type and drainage. Theoretically, there is no limit to how large you can build a raised bed, but available space and watering demands may limit their final size.

SPECIAL PROBLEM: DRAIN FIELD

Digging a trench for drain lines is a lot of work, but it's the best way to remove excess water from low spots with poor natural drainage. Line the trench with clay, concrete tile or plastic drainpipe, which is widely available, inexpensive and less prone to settling and sagging.

Building a Garden Drain

Begin Mark off the low area with garden lime or stake and string, and design your drain line system accordingly.

Next Dig a trench 16–24 in. deep, centered on the lowest spot, at a pitch of 3–4 in. per 100 ft. to a creek, storm drain or dry well.

Third Fill the base of the trench with 2–3 in. of crushed gravel, lay 4 in.-diameter plastic drain pipe with the perforation holes facing down.

Last Cover the pipe with gravel to a depth of 4 in. and tamp down the surface. Backfill the trench.

DRAINAGE PROBLEMS

Poorly drained low spots are usually easy to recognize: Surface water forms pools that remain for hours after a heavy rain, and the ground stays waterlogged and cold for months after better-drained soils have already dried out. Because the water table is too high, spring plantings in poorly drained low spots develop shallow root systems. When the water table drops sharply in hot, dry weather, roots are suddenly parched, and plants wilt. Later, winter cold brings frost-heaving, as freezing and thawing soil pushes the ground.

If drainage testing (see pgs. 18–19) indicates a severe problem, there are two ways to correct it: building up or digging down. Building up involves bringing in new topsoil and improving soil quality and cultivation depth (see pgs. 58–59).

Tube or tile drainage is the time-tested method for piping out excess water from low-lying areas. Bigger gardens will need networks of tube or tile drainpipes in patterns that cover all wet areas, but small gardens can usually get by with a single line of 4 in.-diameter pipe, running down the middle of the garden and out to a ditch, sewer line or creek bed. Take care that the pipes are well-supported to avoid sagging. Covering the pipeline with gravel will ensure that the pipe draws well and drains the field 10 or 15 yds. on either side.

If your low spot has no readily available outlet, dig a dry well—to a pit 3–5 ft. deep and at least 2 ft. in diameter—at the end of your drainpipe. Line the well's sides with plastic and fill with large cobble rock. Cover the top 4 in. with earth.

SPECIAL PROBLEM: HILLSIDE GARDENS

HILLSIDE TERRACES

Steep hillsides are naturally unsuitable for gardening because of constant soil erosion, rapid water runoff and difficulty of cultivation. One of the oldest methods of bringing steeply sloping hillsides under cultivation is terracing, or grading level strips across the slope of the hill. Terracing solves most of the problems of hillside gardening, since it slows erosion, retains water and makes working the ground much easier. Terraces do require effort in their planning, construction and maintenance. If your only available garden area is on a hillside, however, low terraces may be the only solution for bringing it under cultivation.

A well-designed garden terrace, if properly planned and built, can be quite attractive as well. Inspect your hillside site and draw up a terrace plan before you start digging.

The size and setting of the wall should be appropriate to the surrounding landscape. Any shifting of the earth will bring tremendous pressure on the retaining wall, so bear this in mind when specifying materials for construction. Be sure to carefully level your uprights, and allow the footing cement to cure for 2–3 days before adding the horizontal timbers.

Most important, the garden terrace should have ample provision for runoff, so that water doesn't build up behind the terrace. A simple approach is to space the horizontal timbers of the retaining wall slightly apart, so that water will flow through and be carried off down the hillside. Design your terrace's channels and drainage for easy cleaning and upkeep so soil erosion and rapid runoff will never become problems.

A Simple Terraced Garden

First Evaluate your site and sketch out a rough plan.

Then Estimate all materials needed (lumber, gravel, hardware).

Third Cut a trench across the hillside and dig postholes.

Fourth Set 4 x 4 treated posts in footing cement and cure 2–3 days.

Next Bolt horizontal timbers to uprights, spacing them 1/4 in. apart.

Last Backfill with soil and amendments.

CAUTION

Never use preservative-treated lumber in vegetable garden planter beds.

CAUTION

Structural walls over 2 1/2 ft. tall may require building permits.

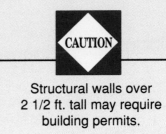

SPECIAL PROBLEM
DRY CLIMATE SOILS

DRY COUNTRY SOILS

In parts of the country with very low rainfall, soils have a whole set of problems rarely seen in the East. The native soils of the desert regions of the western United States are low in organic matter and high in calcium carbonate and calcium sulfate—in a word, alkaline. Alkaline soils absorb water poorly because their surfaces crust over with the salty minerals left after surface evaporation. The ground water is hard, and such soil looks light gray or pale.

Desert soils often compact into a hardpan called *caliche* just below the surface and must be improved before gardening can begin. Replacement of topsoil is an expensive but effective means of making an "instant garden." Irrigation, however, can turn new soil salty and alkaline again. Allow adequate subsurface drainage for the salty water, and flood the soil before spring planting and regularly throughout the summer. This will leach salts out of the soil and balance the pH. High-salt soils can be reclaimed by applying gypsum to the topsoil to replace the sodium or allow it to leach down and away from the plant roots. Or add sulfur, which dissolves the soil lime and restores calcium.

Desert gardeners should try to work with the climate and soil. This means growing native plants, such as cactus, succulents and date palms, and salt-tolerant vegetables, such as beets, spinach and asparagus. Dry country gardeners should also take greater advantage of the rainy winter season to grow overwinter crops. Most importantly, careful watering with drip or ditch irrigation systems will make effective use of the dry country's most precious resource—water.

Till at least 6 in. of new topsoil into native hardpan soil to permit cultivation of non-native plants.

Cacti and succulents will thrive even in rocky desert soil, and they require little care.

65

SPECIAL PROBLEM
THE DEEP SOUTH AND THE NORTHEAST

ACIDIC SOILS

East of the so-called *Lime Line*, in the northern states east of the Ohio and the southern states east of the Mississippi, soils are generally acidic and often require liming. Particularly in high rainfall areas in the deep South, calcium and magnesium are quickly leached out of the soil and need replacing to prevent soil from turning acidic. Living in the North is no guarantee against acidic soil, however. The further north you live, the less evaporation takes place into cooler air, and the less precipitation is needed to leach out the alkaline minerals. The problem of "acid rain" in the industrial northeast further increases the need for liming in that region.

Many of the red soils of the deep South are highly acidic. Even though they are dense with clay, these well-drained upland soils are often depleted of important nutrients. Regular fertilization and liming are the key to gardening success. By raising the pH of red clay soils from 4.5–5.0 to between 5.5–6.5, liming can increase calcium and magnesium as much as 100 percent, and greatly improve soil fertility and plant growth.

Not all of the soils of these regions are acidic, of course. Soils that have developed from limestone, chalk or marl, such as the blackland soil of Alabama or the mountain soil of the Appalachians, rarely need liming. In any case, you'll want to test for acidity (see pgs. 16–17), check with your local Agricultural Extension Service (see pgs. 76–77), and follow the local guidelines for correcting acidity in your soil.

Use extra garden lime to reduce acid in clay soils.

Fertilize rain-depleted soil thoroughly and often.

SPECIAL PROBLEM
GRANITE-BASED SOILS

THE NORTHERN MIDWEST

The soils of the mountainous regions are formed on rocky, shifting terrain, and are often unstable. For example, the parent material of Rocky Mountain soils is granite, which tends to break up and move downhill before completely weathering. The soils of the Appalachian mountain region are residues of limestone, which weathers faster than granite. Many of the soils of the northern Midwest and Alaska are residues of glacial action, have high calcium carbonate and are alkaline. In the Sierra Range, granite-based, high-woodland soils are further depleted by long periods of dryness and continual erosion on steeper slopes.

The mountain soils tend to be rocky and loose, with too much mineral material and not enough organic plant and animal material to hold water and its valuable nutrients. The thin surface soil may lie directly over bedrock. The stony material from which the surface soil has developed is often incompletely weathered, leaving large amounts of rocks, pebbles and sand in the topsoil. Because of the short growing season in the more northern climates, any organic materials also tend to decompose more slowly.

Despite these wide variations in mineral content, pH and organic matter, adding organic compost and other soil amendments will improve garden soils built up from a mineral base. Any native soil mixed in with new topsoil should be thoroughly sieved to remove all rocks larger than coarse sand. To be sure you are using the right soil amendments, check with your local Agricultural Extension Service (see pgs. 76–77) for advice on correcting alkalinity and improving soil texture and drainage.

Add compost to sandy or rocky soil to build up depth and texture.

Fertilize mountain soils well to make up for
depletion of essential nutrients.

SILTY SOILS

In the low-lying regions along streams and rivers, fine sediment material deposited by running water builds up into a silty soil rich in nutrients. Where the river's plain is gently inclined, it swings back and forth to create a meandering river with oxbow curves and backwater lagoons. At high water, the river overflows its banks and deposits sediment across the plain.

Flood plain deposits can vary in texture, but coarser sand and gravel deposits tend to build up along riverbanks. Finer silt and clay materials spread out over the river's flood plain. The Mississippi River flood plain, the largest in North America, is from 25–75 miles wide and over 500 miles long, but every large river from the Sacramento to the Hudson, or the Platte to the Rio Grande, has a flood plain with sediment deposits.

Flood plain soils are usually very rich, but they often need work to be made suitable for gardening. River bottoms and low-lying fields may require drainage, either with tile or tube drainfields or sump-and-pump systems. Following drainage, these soils should be built up with compost or other well-decayed plant material to improve their workability, water retention, and pore space. Sediments carried down the rivers wind up in deltas at the mouths of tributaries, and these delta soils are the richest of all. Flood plain and delta soils can have very different levels of acidity, and soil tests should be made to determine if liming is needed. Consult your local Agricultural Extension Service (see pgs. 76–77) for advice on the particular needs of the soil in your area.

Add compost to give better texture to silty flood-plain soil and help soil drainage.

Apply soil amendments and dropspread any needed fertilizer.

SPECIAL PROBLEM
FOREST SOILS

ACIDIC FOREST SOILS

Much of North America was originally covered by native forests of broadleaved deciduous or evergreen trees, and the organic residue of limbs, leaves and needles from these trees is acidic. Forest soil acidity occurs in many areas—in the maple and birch forests of New England, the oak, hickory and pine forests of the South, and the spruce, fir and redwood forests of the Pacific Coast.

Each forest soil type will vary in terms of acidity, texture, drainage and fertility. Areas of very heavy rainfall will tend to be less acid, for example, and their soils will need less liming as a result. The gray-brown forest soils of the Northeast have a topsoil composed of leaves of maple, beech and oak, and are only slightly acidic. In humid conifer forests, on the other hand, the decomposed needles of spruce and fir produce a highly acid humus with very poor mineral content.

After testing the pH of a forest soil (see pgs. 16–17), select a suitable lime product to sweeten the soil. Don't use quicklime or hydrated lime, the choice for large-scale commercial agriculture; they are too caustic for any garden soil in which you hope to be planting soon. The most popular garden limes—dolomitic limestone or carbonate of lime—are good all-around liming agents that will safely raise pH while adding calcium and magnesium to any garden soil.

Application rates will vary according to the type of soil you have. About 4–5 lbs. for each 100 sq. ft. is sufficient to raise pH value one number in sandy soil. Heavy clay soils are more resistant to lime, and may need as much as 10 lbs. for each 100 sq. ft.

Liming helps correct acidic forest soils, but clay and sand-based soils each need different amounts of lime.

TOOLS AND IMPLEMENTS

Tools for working the soil should be sturdy and comfortable. Try garden tools for fit and feel before buying—they come in many sizes, weights and shapes. Be sure to clean and care for your soil-working tools, oiling them and storing them in a dry place in winter.

Round point shovel

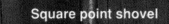
Square point shovel

Barn fork

Spading fork

Weeding tool

Garden hoe

Bow rake

Level
head
rake

Garden shovel

Hand trowel

Transplanting
spade

Getting Garden Advice

Each state's land grant university has a school or college of agriculture. Any of them can refer you to your local Agricultural Extension Service, available in every county, borough or parish in the United States. Nurseries and garden stores can be easily found in the telephone Yellow Pages. If you have further questions concerning federal or state programs in your area, call the U.S. Department of Agriculture Switchboard at 202-447-USDA.

State University Agricultural Extension Services:

ALABAMA Auburn University, Auburn, AL 36849
Telephone: (205) 844-4000

ALASKA University of Alaska, Fairbanks, AK 99701
Telephone: (904) 474-7211

ARIZONA University of Arizona, Tucson, AZ 85721
Telephone: (602) 621-2211

ARKANSAS University of Arkansas, Fayetteville, AR 72701 Telephone: (501) 575-2000

CALIFORNIA University of California, Berkeley, CA 94612 Telephone: (415) 987-0040

University of California, Davis, CA 95616 Telephone: (916) 725-0107

University of California, Riverside, CA 92521
Telephone: (714) 787-1012

COLORADO University of Colorado, Fort Collins, CO 80523 Telephone: (303) 491-1101

CONNECTICUT University of Connecticut, Storrs, CT 06268 Telephone: (303) 282-7300

DELAWARE University of Delaware, Newark, DE 19711
Telephone: (302) 735-8200

DISTRICT OF COLUMBIA University of the District of Columbia, Washington, D.C. 20008 Telephone: (202) 282-7300

FLORIDA University of Florida, Gainesville, FL 32611
Telephone: (904) 392-3261

GEORGIA University of Georgia, Athens, GA 30602
Telephone: (404) 542-3030

HAWAII University of Hawaii, Honolulu, HI 96822
Telephone: (808) 956-8111

IDAHO University of Idaho, Moscow, ID 83843
Telephone: (208) 885-6111

ILLINOIS University of Illinois, Urbana, IL 61801
Telephone: (217) 333-1000

INDIANA Purdue University, Lafayette, IN 47901
Telephone: (317) 494-4600

IOWA Iowa State University, Ames, IA 50010
Telephone: (515) 294-4111

KANSAS Kansas State University, Manhattan, KS 66506 Telephone: (913) 532-6011

KENTUCKY University of Kentucky, Lexington, KY 40506 Telephone: (606) 257-9000

LOUISIANA Louisiana State University, Baton Rouge, LA 70893 Telephone: (504) 388-3202

MAINE University of Maine, Orono, ME 04469
Telephone: (207) 581-1110

MARYLAND University of Maryland, College Park, MD 20742 Telephone: (301) 454-0100

MASSACHUSETTS University of Massachusetts, Amherst, MA 01003 Telephone: (413) 545-0111

MICHIGAN Michigan State University, East Lansing, MI 48824 Telephone: (517) 355-1855

MINNESOTA University of Minnesota, St. Paul, MN 55101 Telephone: (612) 825-2131

MISSISSIPPI Mississippi State University, MS 39762
Telephone: (601) 325-2131

MISSOURI University of Missouri, Columbia, MO 65201 Telephone: (314) 882-2121

MONTANA Montana State University, Bozeman, MT 59715 Telephone: (406) 994-0211

NEBRASKA University of Nebraska, Lincoln, NE 68503
Telephone: (402) 472-7211

NEVADA University of Nevada-Reno, Reno, NV 89507
Telephone: (702) 784-6611

NEW HAMPSHIRE University of New Hampshire, Durham, NH 03824 Telephone: (603) 862-1234

NEW JERSEY Rutgers University, New Brunswick, NJ 08903 Telephone: (201) 932-1766

NEW MEXICO New Mexico State University, Las Cruces, NM 88003 Telephone: (505) 646-0111

NEW YORK Cornell University, Ithaca, NY 14850
Telephone: (607) 256-1000

NORTH CAROLINA   Duke University, Durham, NC
27706 Telephone: (919) 684-2135

NORTH DAKOTA   North Dakota State University,
Fargo, ND 58105 Telephone: (702) 237-9011

OHIO   Ohio State University, Columbus, OH 43210
Telephone: (614) 422-6446

OKLAHOMA   Oklahoma State University, Stillwater,
OK 74708 Telephone: (405) 744-5000

OREGON   Oregon State University, Corvallis, OR 97331
Telephone: (503) 754-0123

PENNSYLVANIA   Pennsylvania State University,
University Park, PA 16802 Telephone: (814) 865-
4700

PUERTO RICO   University of Puerto Rico, Mayaguez,
PR 00708 Telephone: (809) 832-4040

RHODE ISLAND   University of Rhode Island, Kingston,
RI 02881 Telephone: (401) 792-1000

SOUTH CAROLINA   Clemson University, Clemson, SC
29634 Telephone: (803) 656-3311

SOUTH DAKOTA   South Dakota State University,
Brookings, SD 57007 Telephone: (605) 688-4151

TENNESSEE   University of Tennessee, Knoxville, TN
37901 Telephone: (615) 974-0111

TEXAS   Texas A & M University, College Station, TX
77843 Telephone: (409) 845-4747

 Texas Technological University, Lubbock, TX 79409
Telephone: (806) 742-2808

UTAH   Utah State University, Logan, UT 84321
Telephone: (801) 750-1000

VERMONT   University of Vermont, Burlington, VT
05401 Telephone: (802) 656-3131

VIRGINIA   Virginia Polytechnic Institute, Blacksburg,
VA 24061 Telephone: (703) 231-6000

WASHINGTON   Washington State University, Pullman,
WA 99164 Telephone: (509) 335-3564

WEST VIRGINIA   University of West Virginia,
Morgantown, WV 26506 Telephone: (304) 293-0111

WISCONSIN   University of Wisconsin, Madison, WI
53706 Telephone: (608) 262-1234

WYOMING   University of Wyoming, Laramie, WY
82070 Telephone: (307) 766-4133

BOOKS AND ARTICLES FOR FURTHER INFORMATION

Andrews, William A., ed. A Guide to the Study of Soil Ecology. Englewood Cliffs, New Jersey: Prentice-Hall, 1973.

Bartholomew, Mel. Square Foot Gardening. Emmaus, Pennsylvania: Rodale Press, 1981.

"The Basics of Soil Building." Southern Living, August 1987, p. 42.

Brady, Nyle C. The Nature and Properties of Soils. 8th Edition. New York: Macmillan, 1980.

Cook, Jack. "The Importance of Soil Temperature." Horticulture, April 1988, p. 24.

Cox, Jeff. "The Instant Garden." Organic Gardening, December 1987, p. 90.

Crockett, James. Crockett's Tool Shed. Boston: Little, Brown & Co., 1979.

Ferrara, Mike. "Tiller Choices." Organic Gardening, June 1989, p. 47.

Gibson, H. E. "Give Your Lawn Some Air." Flower & Garden Magazine, September-October 1990, p. 56.

"How to Read Your Soil's Mind." Sunset Magazine, May 1988, p. 270.

Logsdon, Gene. The Gardener's Guide to Better Soil. Emmaus, Pennsylvania: Rodale Press, 1975.

Lovejoy, Ann. "The Rush to Compost." Horticulture, February 1989, p. 46.

"Mulch: 1988 Gardening Hero." Sunset Magazine, August 1988, p. 56.

Shirley, Christopher. "Soil Sense." Organic Gardening, June 1989, p. 59.

Sinnes, A. Cort. "Unusual Soil Amendments." Flower and Garden Magazine, March 1988, p. 92.

Sunset Western Garden Book. Menlo Park, California: Sunset Publishing, 1990.

"Very Big Boxes for Hillside Gardening." Sunset Magazine, August 1987, p. 92.

"What About Those New Soil Polymers?" Sunset Magazine, April 1987, p. 252.

INDEX

A Note From NK Lawn and Garden Co.

For more than 100 years, since its founding in Minneapolis, Minnesota, NK Lawn & Garden has provided gardeners with the finest quality seed and other garden products.

We doubt that our leaders, Jesse E. Northrup and Preston King, would recognize their seed company today, but gardeners everywhere in the U.S. still rely on NK Lawn & Garden's knowledge and experience at planting time.

We are pleased to be able to share this practical experience with you through this ongoing series of easy-to-use gardening books.

Here you'll find hundreds of years of gardening experience distilled into easy-to-understand text and step-by-step pictures. Every popular gardening subject is included.

As you use the information in these books, we hope you'll also try our lawn and garden products. They're available at your local garden retailer.

There's nothing more satisfying than a successful, beautiful garden. There's something special about the color of blooming flowers and the flavor of home-grown garden vegetables.

We understand how special you feel about growing things—and NK Lawn & Garden feels the same way, too. After all, we've been a friend to gardeners everywhere since 1884.